GRADE
3 4

SCHOLASTIC

Success With

Multiplication Facts

New York • Toronto • London • Auckland • Sydney
Mexico City • New Delhi • Hong Kong • Buenos Aires

Teaching *Resources*

State Standards Correlations

To find out how this book helps you meet your state's standards, log on to **www.scholastic.com/ssw**

Written by William Earl
Cover design by Ka-Yeon Kim-Li
Interior illustrations by Michael Denman
Interior design by Quack & Company

ISBN 978-0-545-20086-8

17 18 19 20   40   22 21 20

Introduction

Parents and teachers alike will find **Multiplication Facts** to be a valuable learning tool. Students will enjoy completing a wide variety of math activities that are both engaging and educational. Take a look at the Table of Contents. You will feel rewarded providing such a valuable resource for your students. Remember to praise students for their efforts and successes!

Table of Contents

Name _____

A Ray of Fun

An **array** demonstrates a multiplication sentence. The first **factor** tells how many rows there are. The second **factor** tells how many there are in each row. The answer of a multiplication sentence is called the **product**.

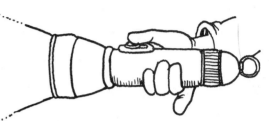

2 x 4 = 8   ○ ○ ○ ○   2 rows
              ○ ○ ○ ○   4 in each row

Write the multiplication sentence for each array.

A.   ○ ○ ○
        ○ ○ ○

3 x 2

B.   ○ ○ ○
        ○ ○ ○
        ○ ○ ○

3 x 3

C.   ○ ○
        ○ ○
        ○ ○
        ○ ○

4 x 2

D.   ○ ○ ○ ○ ○
        ○ ○ ○ ○ ○
        ○ ○ ○ ○ ○

3 x 8

E.   ○ ○ ○

3 x 1

F.   ○ ○ ○
        ○ ○ ○
        ○ ○ ○
        ○ ○ ○

4 x 3

G.   ○ ○ ○ ○ ○ ○
        ○ ○ ○ ○ ○ ○

2 x 6

H.   ○ ○ ○ ○
        ○ ○ ○ ○
        ○ ○ ○ ○

3 x 4

I.   ○ ○ ○ ○ ○ ○
        ○ ○ ○ ○ ○ ○
        ○ ○ ○ ○ ○ ○

3 x 6

J.   ○ ○ ○
        ○ ○ ○
        ○ ○ ○
        ○ ○ ○
        ○ ○ ○

5 x 3

K.   ○
        ○
        ○
        ○
        ○

1 x 5

L.   ○ ○
        ○ ○
        ○ ○
        ○ ○
        ○ ○
        ○ ○
        ○ ○

7 x 2

 It was time for our class photo. The photographer arranged us into four rows. There were six people in each row. How many people in all were in the photo? On another piece of paper, draw an array to solve this problem.

Time to Group

 The multiplication symbol (**x**) can be thought of as meaning "groups of."

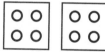

3 "groups of" 4 equals 12
3 x 4 = 12

5 "groups of" 2 equals 10.
5 x 2 = 10

Write the multiplication sentence for each diagram.

A.

4 x 2

B.

3 X 3

C.

5 X 3

D.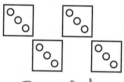

3 X 4

E.

1 X 4

F.

3 x 6

G.

2 X 8

H.

6 X 4

I.

2 X 6

J.

8 X 3

K.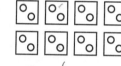

3 X 6

L.

4 X 5

M.

2 X 2

N.

6 x 1

O.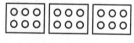

5 X 4

P.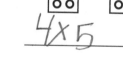

7 X 2

 William has five bags of hamburgers. There are seven hamburgers in each bag. On another piece of paper, show the total number of hamburgers.

Adding Quickly

 The addition sentence 4 + 4 + 4 + 4 + 4 = 20 can be written as a multiplication sentence. Count how many times 4 is being added together. The answer is 5. So, 4 + 4 + 4 + 4 + 4 = 20 can be written as 5 x 4 = 20. Multiplication is a quick way to add.

Write a multiplication sentence for each addition sentence.

A. 5 + 5 + 5 = 15

B. 6 + 6 + 6 + 6 = 24

C. 8 + 8 = 16

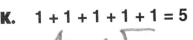

D. 2 + 2 + 2 + 2 = 8

E. 7 + 7 + 7 = 21

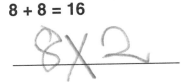

F. 4 + 4 + 4 + 4 = 16

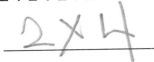

G. 9 + 9 + 9 = 27

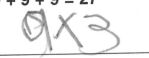

H. 5 + 5 + 5 + 5 + 5 = 25

I. 3 + 3 + 3 + 3 + 3 = 15

J. 10 + 10 + 10 + 10 = 40

K. 1 + 1 + 1 + 1 + 1 = 5

L. 11 + 11 + 11 = 33

M. 8 + 8 + 8 + 8 = 32

N. 0 + 0 + 0 + 0 = 0

O. 12 + 12 + 12 + 12 = 48

P. 9 + 9 + 9 + 9 = 36

Today, we are going to the beach. Mom packed the picnic basket with six sandwiches, six water bottles, six candy bars, and six apples. How many items did she pack in all?

What's My Line?

 5 x 3 = 15 can be demonstrated on a number line.

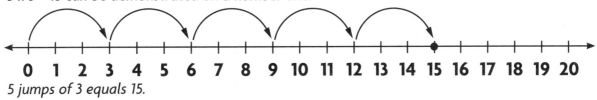

5 jumps of 3 equals 15.

Write the multiplication sentence demonstrated on each number line.

A. 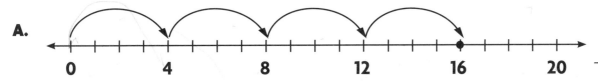  4 x 6

B. 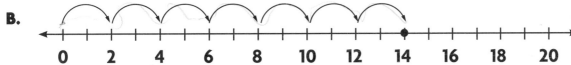  2 x 7

C. 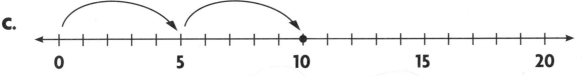  5 x 2

D. 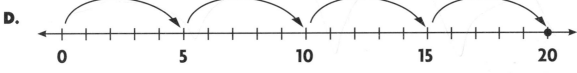  5 x 4

E. 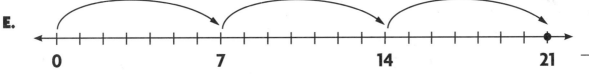  7 x 3

F. 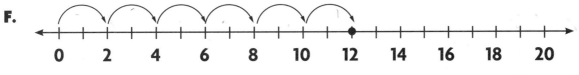  2 x 6

G. 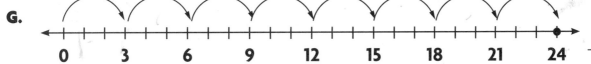 3 x 8

H. 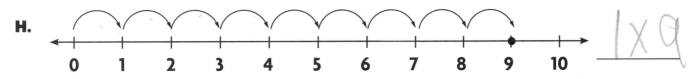 1 x 9

Change It Up

 The order of the factors in a multiplication sentence can change without changing the value of the product. If 2 x 7 is changed to 7 x 2, the product still equals 14.

Change the order of the  factors in each multiplication sentence.

A. 6 x 2 = 12

2X6=12

B. 4 x 8 = 32

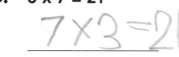 8 X 4 = 32

C. 3 x 9 = 27

9X3=87

D. 3 x 7 = 21

7X3=21

E. 5 x 9 = 45

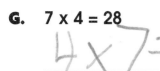 9X5=45

F. 6 x 7 = 42

 7X6=42

G. 7 x 4 = 28

4X7=28

H. 3 x 12 = 36

 12X3=3

I. 9 x 8 = 72

8X9=72

J. 6 x 5 = 30

 5X6=30

K. 4 x 10 = 40

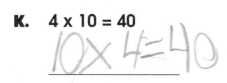 10X4=40

L. 9 x 7 = 63

 7X0=63

M. 2 x 11 = 22

11X2=22

N. 12 x 11 = 132

11X12=13

Name _____ **Identifying fact families**

Family Fun

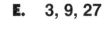 *Multiplication is the opposite of division. The product and factors can be used to write division sentences. The multiplication and division sentences are called a **fact family**.*

2 x 6 = 12 (2 groups of 6)   12 ÷ 6 = 2 (12 divided into 6 equal groups)

6 x 2 = 12 (6 groups of 2)   12 ÷ 2 = 6 (12 divided into 2 equal groups)

Write two multiplication and two division sentences for each set of numbers.

A. 2, 3, 6

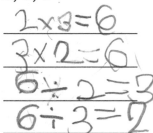
2 x 3 = 6
3 x 2 = 6
6 ÷ 2 = 3
6 ÷ 3 = 2

B. 2, 8, 16

_____
_____
_____
_____

C. 4, 5, 20

_____
_____
_____
_____

D. 3, 5, 15

_____
_____
_____
_____

E. 3, 9, 27

_____
_____
_____
_____

F. 3, 12, 36

_____
_____
_____
_____

G. 5, 6, 30

_____
_____
_____
_____

H. 6, 7, 42

_____
_____
_____
_____

I. 4, 8, 32

_____
_____
_____
_____

 Ramone has 33 marbles. He keeps an equal number of marbles in each of 3 bags. How many marbles are in each bag? On another piece of paper, write a number sentence to solve this problem. Then write the set of numbers in this fact family.

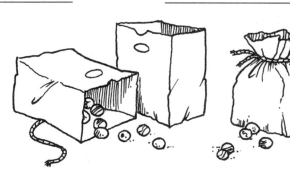

Grin and Count It

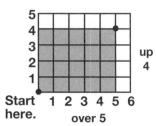

A multiplication sentence can be diagrammed on a **coordinate grid**. *To show 5 x 4, use the factors as the ordered pair (5, 4). Then go over 5 and up 4 on the grid, and mark the point where the lines intersect to make a rectangle. Finally, count all the squares in the rectangle.*

First, write the ordered pair for each set of multiplication factors. Then mark the intersecting point for the ordered pair on the grid. Color and count each square in the rectangle. Fill in the blank with the total number of squares that are in the rectangle.

A.

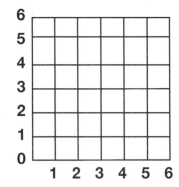

3 x **4** = ( )

Total squares = _____

B.

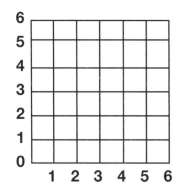

2 x **4** = ( )

Total squares = _____

C.

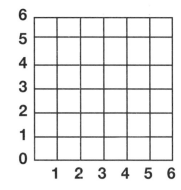

3 x **2** = ( )

Total squares = _____

D.

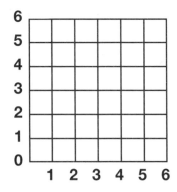

1 x **5** = ( )

Total squares = _____

E.

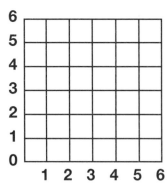

5 x **2** = ( )

Total squares = _____

F.

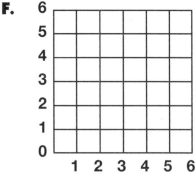

3 x **6** = ( )

Total squares = _____

 Before Amanda opened her present, she measured it. It was eight inches long and four inches wide. On another piece of paper, draw a coordinate grid to show the total number of square inches Amanda's present measured.

Name _____

Find the Patterns

What is the pattern for the numbers 0, 2, 4, 6, 8, 10, 12, 14, 16, 18?
The pattern shows multiples of 2.

Complete each pattern.

A. 3, 6, 9, 12, _____, _____, _____, _____, _____

B. 4, 8, 12, 16, _____, _____, _____, _____, _____

C. 1, 2, 3, 4, _____, _____, _____, _____, _____

D. 7, 14, 21, _____, _____, _____, _____, _____

E. 10, 20, 30, _____, _____, _____, _____, _____

F. _____, 18, 27, _____, _____, _____, _____

G. 6, 12, _____, _____, 30, _____, _____, _____

H. _____, 22, _____, 44, _____, _____, 77

I. 5, 10, 15, _____, _____, _____, _____, _____

J. 8, _____, 24, _____, 40, _____, _____, _____

K. 10, 12, 14, _____, _____, _____, 22, _____, _____

L. _____, 24, _____, 48, 60, _____, _____, _____, _____

Sam ran every afternoon last week. On Sunday, he ran 3 miles. On Monday, he ran 6 miles.
On Wednesday, he ran 12 miles. How many miles do you think he ran on Tuesday?

Code Zero! Code One!

 When a number is multiplied by 0, the product is always 0.
When a number is multiplied by 1, the product is always the number being multiplied.

Multiply. Shade all products of 0 yellow. Shade all other products green.

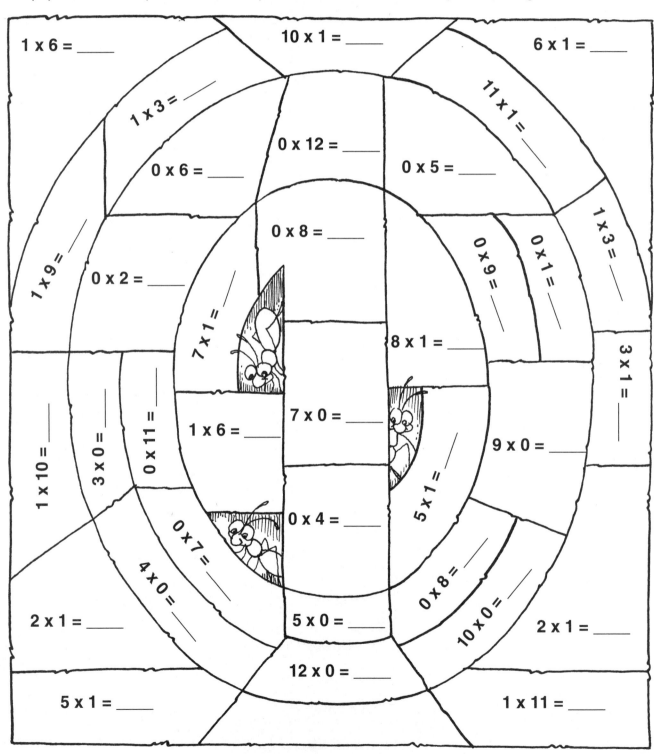

1 x 6 = _____

10 x 1 = _____

6 x 1 = _____

1 x 3 = _____

11 x 1 = _____

0 x 12 = _____

0 x 6 = _____

0 x 5 = _____

0 x 8 = _____

0 x 9 = _____

0 x 1 = _____

1 x 3 = _____

1 x 9 = _____

0 x 2 = _____

7 x 1 = _____

8 x 1 = _____

3 x 1 = _____

7 x 0 = _____

1 x 6 = _____

9 x 0 = _____

1 x 10 = _____

3 x 0 = _____

0 x 11 = _____

0 x 4 = _____

5 x 1 = _____

0 x 7 = _____

4 x 0 = _____

0 x 8 = _____

2 x 1 = _____

5 x 0 = _____

10 x 0 = _____

2 x 1 = _____

12 x 0 = _____

5 x 1 = _____

1 x 11 = _____

Two, Four, Six, Eight, Who Do We Appreciate?

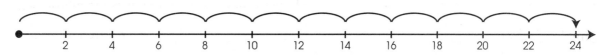 *When multiplying by 2, skip count by 2, or think of number line jumping!*

Multiply.

A. 2 x 3 = _____   2 x 8 = _____   11 x 2 = _____   2 x 7 = _____

B. 8 x 2 = _____   4 x 2 = _____   2 x 2 = _____   2 x 4 = _____

C. 12 x 2 = _____   5 x 2 = _____   10 x 2 = _____   2 x 12 = _____

D. 9 x 2 = _____   2 x 1 = _____   2 x 10 = _____   7 x 2 = _____

E. 2 x 0 = _____   2 x 6 = _____   3 x 2 = _____   0 x 2 = _____

F. 2 x 5 = _____   2 x 9 = _____

G. 6 x 2 = _____   1 x 2 = _____

H. 2 x 11 = _____   2 x 2 = _____

 On another piece of paper, write a rhyme to go with each multiplication fact for 2.
Examples: "2 x 4 = 8, I love math. Can you relate?" Or, "2 x 4 = 8, I've got to go, and shut the gate!"

Aim for the Stars

Color each cloud with a correct multiplication sentence to show the path to the space station.

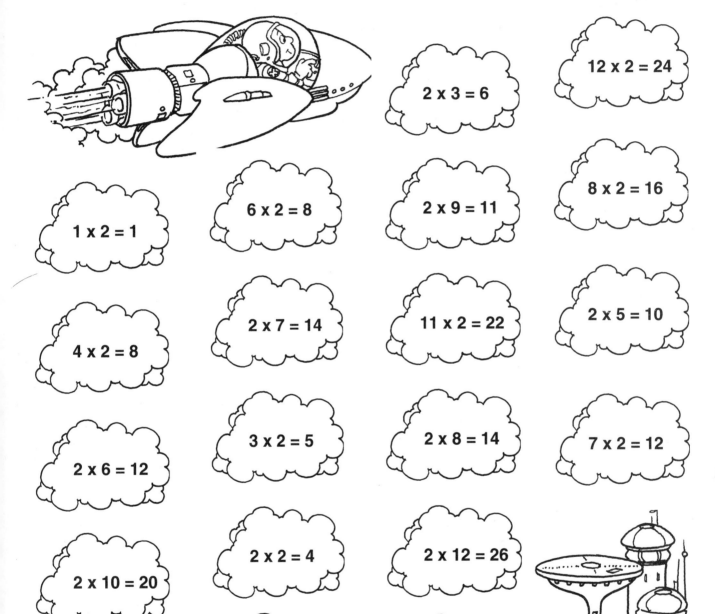

2 x 3 = 6

12 x 2 = 24

1 x 2 = 1

6 x 2 = 8

2 x 9 = 11

8 x 2 = 16

4 x 2 = 8

2 x 7 = 14

11 x 2 = 22

2 x 5 = 10

2 x 6 = 12

3 x 2 = 5

2 x 8 = 14

7 x 2 = 12

2 x 10 = 20

2 x 2 = 4

2 x 12 = 26

9 x 2 = 18

2 x 1 = 2

Eight pilots each flew a plane across the Atlantic Ocean. Each pilot invited one passenger to fly with her. How many people in all flew across the Atlantic Ocean?

The Three Factory

Multiply.

START →

3 x 12 = 36
3 x 4 = 12
3 x 9 = 27

3 x 0 = 0
9 x 3 = 27
6 x 3 = 18

2 x 3 = 6
8 x 3 = 24
4 x 3 = 12

5 x 3 = 15
3 x 3 = 9
3 x 1 = 3

3 x 10 = 30
3 x 6 = 18

12 x 3 = 36
3 x 8 = 24
1 x 3 = 3
7 x 3 = 21
3 x 5 = 15
10 x 3 = 30

3 x 11 = 33
0 x 3 = 0
3 x 4 = 18
6 x 3 = ___
8 x 3 = ___
3 x 12 = ___

3 x 3 = ___
3 x 7 = ___
3 x 2 = ___
10 x 3 = ___
3 x 9 = ___
11 x 3 = ___

The Three Factory paints one stack of boxes every three minutes. How many minutes does it take the factory to paint nine stacks of boxes?

Name _____

A Positive Answer

What should you say if you are asked, "Do you want to learn the 3s?"

To find out, look at each problem below. If the product is correct, color the space green. If the product is incorrect, color the space yellow.

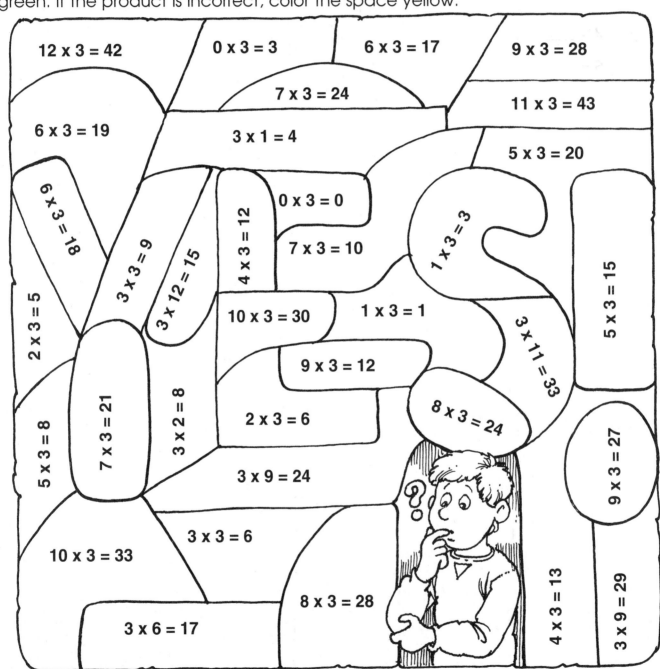

12 x 3 = 42

0 x 3 = 3

6 x 3 = 17

9 x 3 = 28

7 x 3 = 24

11 x 3 = 43

6 x 3 = 19

3 x 1 = 4

5 x 3 = 20

6 x 3 = 18

0 x 3 = 0

1 x 3 = 3

5 x 3 = 15

3 x 3 = 9

3 x 12 = 15

4 x 3 = 12

7 x 3 = 10

2 x 3 = 5

10 x 3 = 30

1 x 3 = 1

3 x 11 = 33

9 x 3 = 12

5 x 3 = 8

7 x 3 = 21

3 x 2 = 8

2 x 3 = 6

8 x 3 = 24

9 x 3 = 27

3 x 9 = 24

3 x 3 = 6

10 x 3 = 33

8 x 3 = 28

4 x 3 = 13

3 x 9 = 29

3 x 6 = 17

How many letters are in the answer to the puzzle? If you wrote this word ten times, how many letters would you write altogether?

Puzzling Facts

Multiply. Write the number word for each product in the puzzle. Don't forget the hyphens!

Across

2. 4 x 9 = _____

4. 4 x 5 = _____

7. 4 x 3 = _____

8. 4 x 7 = _____

9. 4 x 10 = _____

11. 4 x 0 = _____

12. 4 x 11 = _____

Down

1. 4 x 4 = _____

2. 4 x 8 = _____

3. 4 x 12 = _____

5. 4 x 2 = _____

6. 4 x 6 = _____

10. 4 x 1 = _____

 Tracy was missing 4 buttons on 11 different shirts. How many buttons does she need to fix all the shirts?

Fantastic Four

Don't you just adore the factor 4?

To answer this question, multiply. Then use the code to write the letter of each multiplication sentence on the blank above its product.

A. 4 x 10 =	**I.** 4 x 0 =	**O.** 4 x 7 =	**T.** 4 x 8 =
D. 4 x 4 =	**M.** 4 x 2 =	**R.** 4 x 6 =	**Y.** 4 x 9 =
E. 4 x 11 =	**N.** 4 x 5 =	**S.** 4 x 3 =	**!.** 4 x 12 =

___ ___ ___ ___   ___   ___ ___ ___ ___ ___   ___ ___
36   44   12   48   0   40   16   28   24   44   0   32

___ ___ ___ ___   ___ ___ ___   ___ ___ ___ ___ ___
 8   28   24   44   40   20   16   8   28   24   44   48

 On another piece of paper, write a message to a friend. Make a code using the multiplication facts for 4. Have your friend use the code to read the message.

We Can Make Fives Come Alive and Thrive!

What letter stands for "math" and "multiplication"?

To find out, complete each problem. Connect the dots in order from least to greatest.

5 x 2 =
_____

5 x 4 =
_____

5 x 1 = •

• 5 x 5 =
_____

5 x 11 =
_____
•

5 x 8 =
_____
•

•
5 x 3 =
_____

5 x 10 =
_____

5 x 9 =
_____

5 x 0 = •——•   •   •   •   • 5 x 6 =
_____   5 x 12 =   5 x 7 =   _____
_____   _____

 There are five children in line to buy ice cream cones. If each child buys a cone with three scoops of ice cream, how many total scoops of ice cream will the store sell?

How Many Can You Find?

Complete each multiplication sentence. Then circle each answer in the picture.

A. $2 \times 5 = $ _____

B. $5 \times$ _____ $= 5$

C. _____ $\times 5 = 35$

D. $10 \times 5 = $ _____

E. _____ $\times 5 = 60$

F. $5 \times 6 = $ _____

G. _____ $\times 5 = 55$

H. $5 \times 3 = $ _____

I. $8 \times 5 = $ _____

J. _____ $\times 5 = 45$

K. $2 \times$ _____ $= 10$

L. _____ $\times 5 = 25$

M. $7 \times 5 = $ _____

N. $5 \times 12 = $ _____

O. $5 \times$ _____ $= 20$

 Squeaky Squirrel lived in a tree with 4 squirrel friends. If each squirrel collected 12 nuts, how many nuts altogether did the squirrels collect?

Follow the Path

➡ *Let's review! The multiplication symbol (x) means "groups of."*

Multiply. Then follow the path from each multiplication sentence to its product.

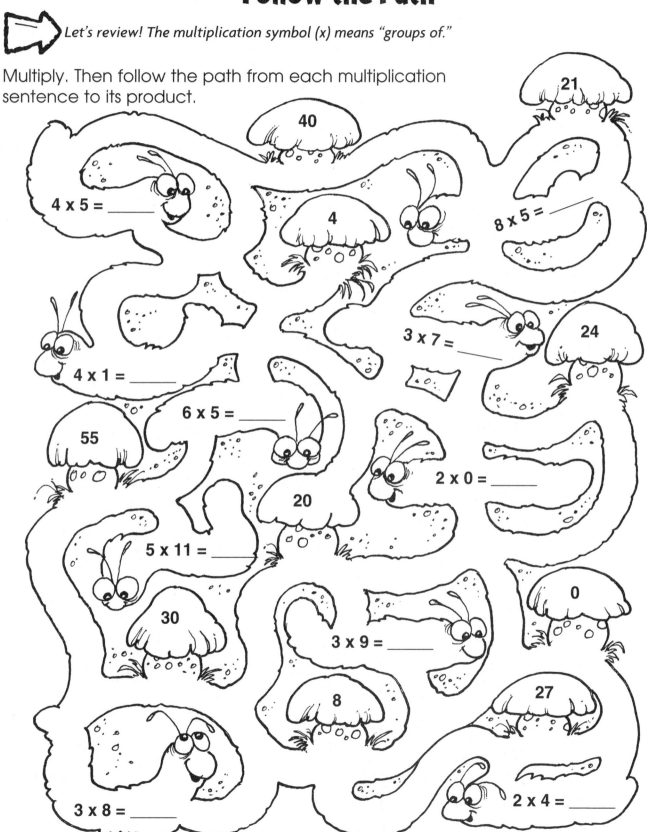

4 x 5 = _____

8 x 5 = _____

3 x 7 = _____

4 x 1 = _____

6 x 5 = _____

2 x 0 = _____

5 x 11 = _____

3 x 9 = _____

3 x 8 = _____

2 x 4 = _____

40

21

4

24

55

20

0

30

8

27

Riddle and Review

 Let's review some more! The numbers being multiplied are called factors. The answer is called the product.

Why did the math teacher choose multiplication to help his class grow?

To find out, multiply. Use the code to write the letter of each multiplication sentence on the blank above its product.

A.	3 x 12 =	**H.**	2 x 9 =	**O.**	3 x 7 =	**U.**	2 x 12 =
B.	5 x 10 =	**I.**	4 x 7 =	**P.**	1 x 0 =	**W.**	5 x 5 =
D.	2 x 8 =	**L.**	5 x 6 =	**R.**	2 x 11 =	**Y.**	4 x 12 =
E.	4 x 11 =	**M.**	4 x 8 =	**S.**	5 x 7 =	**!.**	3 x 3 =
G.	2 x 6 =	**N.**	3 x 9 =	**T.**	5 x 9 =		

___ ___   ___ ___ ___ ___   ___ ___ ___   ___ ___ ___ ___ ___
35 21   45 18 36 45   18 28 35   12 22 21 24   0

___ ___ ___ ___ ___   ___ ___ ___ ___ ___   ___ ___
25 21 24 30 16   50 44 12 28 27   45 21

___ ___ ___   ___ ___ ___ ___ ___ ___   ___ ___ ___
12 44 45   30 36 22 12 44 22   36 27 16

___ ___ ___ ___ ___ ___ ___ ___ ___
32 24 30 45 28   0 30 48   9

 On the field trip to the Science Museum, Mr. Weaver divided his class into six groups. Mrs. Moore divided her class into five groups. Each group had four students. How many students are in each class? Which class has more students? Solve the problem on another piece of paper.

Quality Math

How can you be sure that multiplication is quality math?

To find out, multiply. Then use the code to write the letter of each multiplication sentence on the blank above its product.

A.	2 x 11 =	**H.**	5 x 12 =	**S.**	5 x 9 =
B.	1 x 7 =	**I.**	2 x 6 =	**T.**	5 x 7 =
C.	2 x 9 =	**M.**	4 x 11 =	**U.**	4 x 7 =
D.	4 x 12 =	**O.**	3 x 7 =	**V.**	3 x 11 =
E.	4 x 9 =	**P.**	5 x 8 =	**Y.**	3 x 8 =
G.	4 x 4 =	**R.**	5 x 6 =	**!.**	5 x 0 =

| 7 | 36 | 18 | 22 | 28 | 45 | 36 | | 24 | 21 | 28 | | 60 | 22 | 33 | 36 |

| 35 | 60 | 36 | | 30 | 12 | 16 | 60 | 35 | | 40 | 30 | 21 | 48 | 28 | 18 | 35 |

| 36 | 33 | 36 | 30 | 24 | | 35 | 12 | 44 | 36 | 0 |

Andrew bought nine packages of crackers. There were four crackers in each package. How many crackers did he buy altogether?

Mark Becomes the Teacher!

Write each product to complete the story.

One day in September, our teacher, Mr. Schnoodles, said he needed to step out of the classroom for a brief moment. After _____ seconds, I realized it was
<u>5 x 11</u>
_____ o'clock and time for recess. Suddenly, our classroom phone started to ring.
<u>2 x 5</u>
It rang _____ times before I finally picked up the receiver. It was Mr. Schnoodles!
<u>4 x 8</u>
Mr. Schnoodles asked me to take charge of the class for _____ hours, so that he
<u>3 x 2</u>
could help the principal in the office! I thought this over for _____ seconds and
<u>3 x 6</u>
then replied, "Okay! No problem!"

I walked to the front of the classroom on September _____ to tell the other
<u>3 x 9</u>
_____ students that I was now in charge. The _____ other students cheered,
<u>4 x 6</u>   <u>3 x 8</u>
"Mark, Mark, he's our man, if anyone can teach us, we know he can!" I then announced that since we were _____ minutes late for our _____-minute recess, I
<u>4 x 1</u>   <u>5 x 3</u>
would extend the recess for _____ minutes. Again, the students cheered!
<u>4 x 11</u>
After recess, it was time for _____ minutes of math. I decided we would play
<u>5 x 10</u>
Head's Up, _____-Up. "That's a form of math, isn't it?" I thought. No one
<u>1 x 7</u>
complained. However, after _____ times of trying to guess who pushed down
<u>4 x 12</u>
your thumb, we all became a little bored! I tried to remember what Mr. Schnoodles did when all _____ of his students became bored. I then realized that
<u>5 x 5</u>
in the _____ days we had been in school, we had never been bored! He always
<u>2 x 9</u>
had planned at least _____ fun activities every day. My favorite math activity
<u>3 x 7</u>
was to see how many multiplication problems I could solve in _____ seconds. So,
<u>5 x 12</u>
I gave every student, including myself, a multiplication worksheet with _____
<u>4 x 10</u>
problems. "Let's see how many problems we can solve in _____ seconds," I
<u>12 x 5</u>
instructed. "Ready, set, go!"

Suddenly at _____ minutes after _____, I heard my alarm. It was just a dream!
<u>5 x 3</u>   <u>7 x 1</u>

Mathematics Fireworks

Multiply. On another piece of paper, find the sum of the products of each star trail. Then use the key to color each star to match its star trail sum.

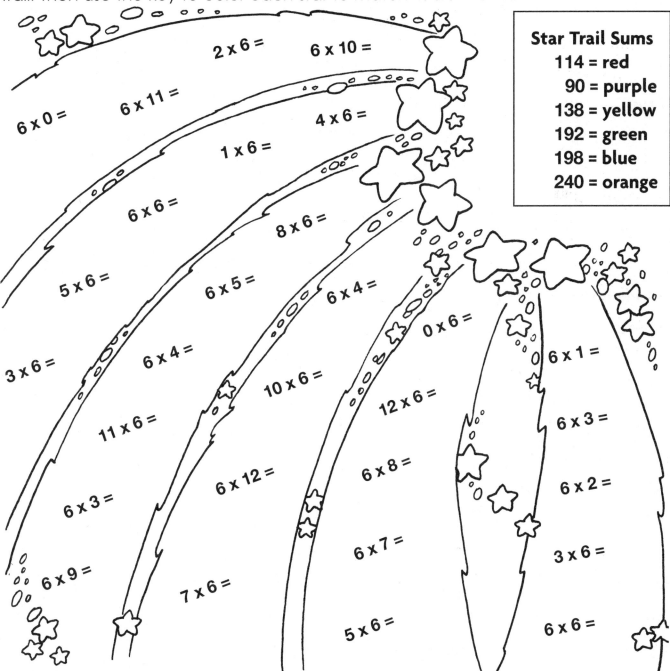

Star Trail Sums
114 = red
90 = purple
138 = yellow
192 = green
198 = blue
240 = orange

💡 **Emma counted the fireworks she watched on the Fourth of July. She counted 6 different fireworks every 15 minutes. The firework show lasted 2 hours. How many fireworks did Emma see?**

Name _____

Can You Crack the Code?

≈	●	⊠	☺	⇐	★	✓	∇	↙	◗	☺	ϒ	⌘
0	1	2	3	4	5	6	7	8	9	10	11	12

Using the above code, write a multiplication sentence for each message.

A. ☺ x ✓ = ●↙   **B.** ✓ x ★ = ☺≈   **C.** ◗ x ✓ = ★⇐   **D.** ϒ x ✓ = ✓✓

_____   _____   _____   _____

E. ✓ x ⊠ = ●⊠   **F.** ⇐ x ✓ = ⊠⇐   **G.** ↙ x ✓ = ⇐↙   **H.** ∇ x ✓ = ⇐⊠

_____   _____   _____   _____

I. ✓ x ⌘ = ∇⊠   **J.** ≈ x ✓ = ≈   **K.** ● x ✓ = ✓   **L.** ☺ x ✓ = ✓≈

_____   _____   _____   _____

Multiply. Then use the above code to write each multiplication sentence.

M. 5 x 6 = _____   **N.** 6 x 7 = _____   **O.** 6 x 9 = _____   **P.** 6 x 3 = _____

_____   _____   _____   _____

Q. 6 x 8 = _____   **R.** 6 x 6 = _____   **S.** 12 x 6 = _____   **T.** 6 x 10 =

_____   _____   _____   _____

Abby wrote the same message to 6 different friends. She made a code using flower symbols for each of the 12 letters in her message. How many total flower symbols did she write?

The "Seven" Statues

Multiply.

$7 \times 2 =$ _____

$1 \times 7 =$ _____

$8 \times 7 =$ _____

$12 \times 7 =$ _____

$7 \times 5 =$ _____

$7 \times 9 =$ _____

$7 \times 10 =$ _____

$0 \times 7 =$ _____

$7 \times 11 =$ _____

$7 \times 12 =$ _____

$7 \times 7 =$ _____

$7 \times 8 =$ _____

$3 \times 7 =$ _____

$9 \times 7 =$ _____

$7 \times 4 =$ _____

$11 \times 7 =$ _____

$6 \times 7 =$ _____

Maurice was hired to build seven statues in front of City Hall. He calculated that each statue would take him six months to finish. The statues need to be completed before the Music Festival that is scheduled to take place in exactly two years. How many months will it take Maurice to complete the statues? Will Maurice have enough time?

Flying Sevens

Multiply.

7 x 9 = ____

11 x 7 = ____

6 x 7 = ____

7 x 4 = ____

3 x 7 = ____

7 x 7 = ____

7 x 10 = ____

7 x 0 = ____

5 x 7 = ____

7 x 12 = ____

7 x 2 = ____

4 x 7 = ____

7 x 11 = ____

1 x 7 = ____

0 x 7 = ____

7 x 8 = ____

2 x 7 = ____

7 x 1 = ____

7 x 6 = ____

8 x 7 = ____

9 x 7 = ____

10 x 7 = ____

12 x 7 = ____

7 x 3 = ____

7 x 5 = ____

Cassandra's space mission is to orbit Earth seven times, as quickly as she can a total of seven times. How many times altogether will she orbit Earth?

The Ultimate Eight Track

Use a stopwatch to time how long it takes to multiply around the track.

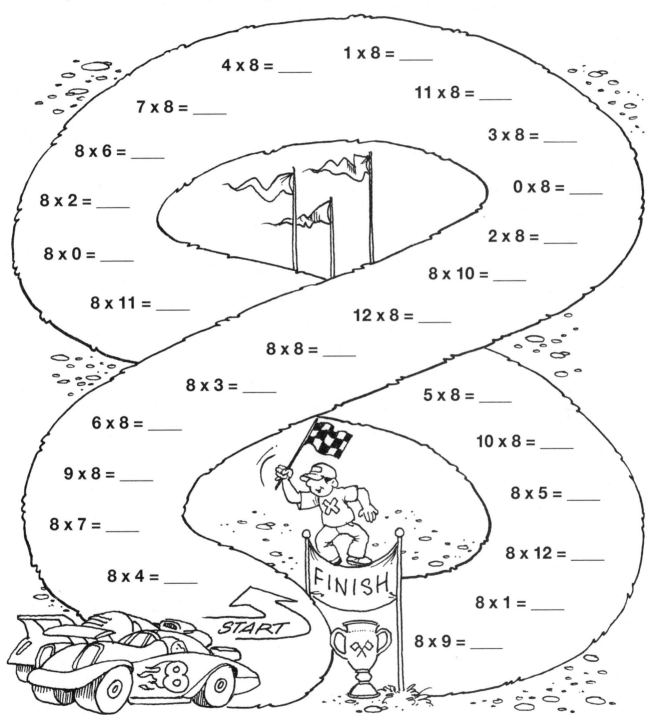

$4 \times 8 =$ ____

$1 \times 8 =$ ____

$7 \times 8 =$ ____

$11 \times 8 =$ ____

$8 \times 6 =$ ____

$3 \times 8 =$ ____

$8 \times 2 =$ ____

$0 \times 8 =$ ____

$8 \times 0 =$ ____

$2 \times 8 =$ ____

$8 \times 11 =$ ____

$8 \times 10 =$ ____

$12 \times 8 =$ ____

$8 \times 8 =$ ____

$8 \times 3 =$ ____

$5 \times 8 =$ ____

$6 \times 8 =$ ____

$10 \times 8 =$ ____

$9 \times 8 =$ ____

$8 \times 5 =$ ____

$8 \times 7 =$ ____

$8 \times 12 =$ ____

$8 \times 4 =$ ____

$8 \times 1 =$ ____

$8 \times 9 =$ ____

 Racing Ricardo rapidly raced 8 times around the Eight Track. It took him 12 seconds to race one time around the track. How many seconds did it take him to complete the race?

A Product Search

Multiply. Then circle the number word for each product in the puzzle. The words will go forward, backward, up, down, and diagonally. Be careful; some products appear more than once!

A. 8 x 2 = _____   4 x 8 = _____   8 x 4 = _____   10 x 8 = _____

B. 0 x 8 = _____   5 x 8 = _____   8 x 6 = _____   9 x 8 = _____

C. 8 x 1 = _____   8 x 3 = _____   2 x 8 = _____   11 x 8 = _____

D. 1 x 8 = _____   8 x 12 = _____   3 x 8 = _____   6 x 8 = _____

E. 8 x 5 = _____   8 x 8 = _____   8 x 0 = _____

F. 8 x 9 = _____   8 x 7 = _____   8 x 8 = _____

F	C	E	L	I	M	R	U	O	F	–	Y	T	N	E	W	T
O	O	F	O	R	T	Y	–	E	I	G	H	T	I	F	E	H
R	N	I	S	I	X	C	B	I	F	N	E	E	T	X	I	S
T	S	F	I	J	W	E	I	G	H	T	S	T	S	O	G	I
Y	I	T	X	T	F	H	R	H	S	Z	E	R	O	W	H	L
–	O	Y	T	W	V	O	S	T	I	U	V	W	V	T	T	R
E	W	–	E	E	U	I	R	Y	X	B	E	X	Y	–	Y	U
I	T	S	E	N	S	M	L	T	T	C	N	A	Z	Y	–	O
G	–	I	N	T	H	I	R	T	Y	–	T	W	O	T	E	F
H	Y	X	X	Y	Z	N	P	Q	–	D	Y	J	N	N	I	–
T	T	Y	Y	–	T	E	R	A	F	E	–	F	Q	E	G	Y
H	R	D	T	F	D	U	R	Z	O	G	T	K	O	V	H	T
G	I	H	R	O	F	V	Y	O	U	F	W	L	U	E	T	X
I	H	Y	O	U	E	W	X	C	R	H	O	M	W	S	T	I
E	T	L	F	R	N	I	N	E	T	Y	–	S	I	X	Z	S

Is There a Pattern?

Is there a pattern of the products when multiplying by 9? Yes! The sum of each product equals 9! There are two exceptions. One exception is 11 x 9 = 99; then each number in the product is 9! What is the other exception?

$0 \times 9 =$ 0
$1 \times 9 =$ 9
$2 \times 9 =$ 1 8
$3 \times 9 =$ 2 7
$4 \times 9 =$ 3 6
$5 \times 9 =$ 4 5
$6 \times 9 =$ 5 4
$7 \times 9 =$ 6 3
$8 \times 9 =$ 7 2
$9 \times 9 =$ 8 1
$10 \times 9 =$ 9 0

THERE SEEMS TO BE A PATTERN HERE.

Unscramble the number word for each product of the following 9s multiplication facts. Then write the 9s fact next to the number word.

A. ENO DDHNUER TEHGI _____

B. NYTENI-NNEI _____

C. TINENY _____

D. YITHEG-NOE _____

E. EENTVSY-WOT _____

F. YXTIS-ERETH _____

G. TFFIY-RUOF _____

H. YTRFO-EVIF _____

I. YHRTTI-XIS _____

On another piece of paper, write the pattern for the above facts. Then write the remaining 9s multiplication facts to complete the pattern.

Name _____

Cross-Number Puzzle

Multiply. Write the number word for each product in the puzzle. Don't forget the hyphens!

Across

1. 9 x 5 = ____
5. 1 x 9 = ____
7. 9 x 12 = ____
9. 4 x 9 = ____
10. 9 x 10 = ____
12. 2 x 9 = ____
13. 9 x 11 = ____

Down

2. 9 x 3 = ____
3. 6 x 9 = ____
4. 0 x 9 = ____
6. 9 x 8 = ____
8. 7 x 9 = ____
11. 9 x 9 = ____

Justin just finished putting together a puzzle of a castle and wants to know how many pieces are in the puzzle. He knows he put together nine pieces every five minutes. If Justin worked for one hour, how many pieces does the puzzle have?

Geometric Multiplication

Multiply. Color each triangle with an even product orange. Color each triangle with an odd product blue.

8 x 6 = ____	9 x 4 = ____	8 x 9 = ____	8 x 12 = ____
7 x 9 = ____	7 x 7 = ____	9 x 3 = ____	9 x 11 = ____
7 x 7 = ____	4 x 6 = ____	8 x 7 = ____	1 x 7 = ____
8 x 8 = ____	9 x 5 = ____	5 x 7 = ____	8 x 10 = ____
6 x 9 = ____	9 x 9 = ____	7 x 3 = ____	6 x 6 = ____
7 x 11 = ____	5 x 8 = ____	6 x 3 = ____	9 x 7 = ____
1 x 9 = ____	5 x 9 = ____	7 x 5 = ____	3 x 9 = ____
7 x 10 = ____	7 x 6 = ____	9 x 8 = ____	6 x 12 = ____

 Maria was decorating a picture frame for her friend's birthday. She chose seven different-sized, diamond-shaped tiles to glue around the frame. There was enough room to glue four colors of each size of tile. How many tiles did she use altogether to decorate the frame? On another piece of paper, solve this problem and draw a picture of what the frame might look like.

Up, Up, and Away

Color each box that contains a multiplication sentence that is correct. Cross out each incorrect product and write the correct product.

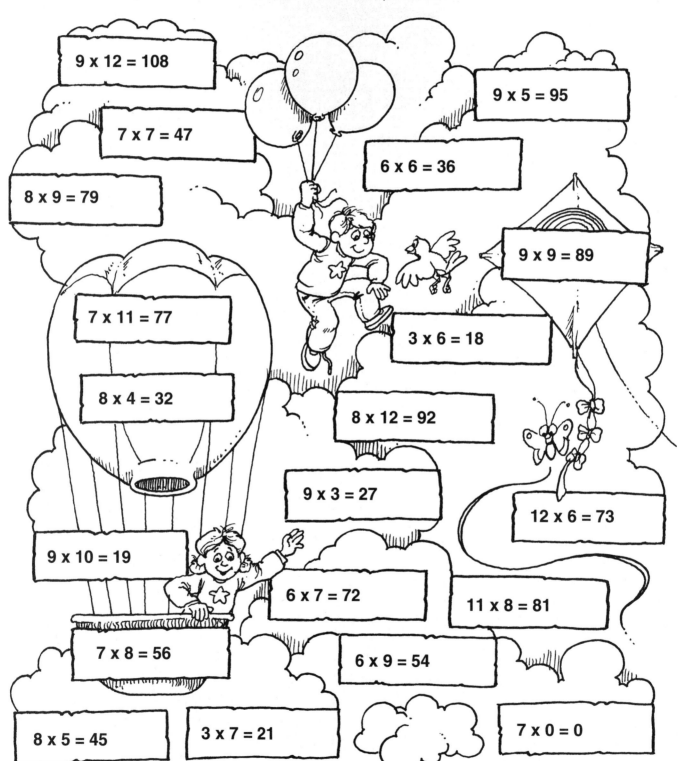

9 x 12 = 108

9 x 5 = 95

7 x 7 = 47

6 x 6 = 36

8 x 9 = 79

9 x 9 = 89

7 x 11 = 77

3 x 6 = 18

8 x 4 = 32

8 x 12 = 92

9 x 3 = 27

12 x 6 = 73

9 x 10 = 19

6 x 7 = 72

11 x 8 = 81

7 x 8 = 56

6 x 9 = 54

8 x 5 = 45

3 x 7 = 21

7 x 0 = 0

That's Some Chatter

Write each product to complete the story.

Once upon a time there was a dog named Chatter who

learned how to talk! That's right! This incredible canine could say

_____ words with one breath! Chatter was so good at talking that
3 x 9

one day he answered the phone _____ times. All _____ people who called
2 x 12   8 x 3

thought it was Chatter's master talking!

On January _____, this news was leaked to the media. Instantly, _____
4 x 5   7 x 4

television stations wanted to schedule a _____-minute interview with Chatter.
3 x 3

Chatter became very famous, to say the least! He was so popular that he was

even asked to be on a national television show! He was offered _____ dog
8 x 9

bones for each guest appearance he would make! After talking for _____ days
4 x 4

with his owners, Chatter declined. It was estimated that _____ million people
6 x 6

were extremely disappointed with this decision.

Chatter was very happy. He had every comfort a canine could want. Believe

it or not, he had _____ air-conditioned doghouses and _____ people to pet
5 x 12   3 x 10

him, including _____ highly trained tummy rubbers. Chatter had even received
5 x 1

table scraps from _____ very famous people!
3 x 7

One day, on an early morning walk with his private trainer, Chatter's

conversation ceased! Chatter had held his own in conversation for _____ days.
7 x 8

And now, just as quickly as it came, it ended. Chatter was relieved. Now he could

be a normal dog again like all his _____ barking friends in the neighborhood.
9 x 10

Name _____

A Learning Lesson

What's the best way to learn multiplication?

To find out, multiply. Then use the code to write the letter of each multiplication sentence on the blank above its product.

A. 1 x 6 =	**G.** 4 x 4 =	**M.** 6 x 12 =	**S.** 9 x 3 =
C. 5 x 9 =	**H.** 8 x 8 =	**N.** 7 x 7 =	**T.** 8 x 12 =
D. 7 x 8 =	**I.** 10 x 9 =	**O.** 2 x 6 =	**U.** 9 x 9 =
E. 5 x 5 =	**K.** 3 x 12 =	**P.** 9 x 2 =	**Y.** 9 x 12 =
F. 4 x 6 =	**L.** 7 x 12 =	**R.** 12 x 5 =	**!.** 0 x 3 =

___ ___ ___ ___   ___ ___ ___   ___ ___ ___ ___ ___
24 90 49 56   96 64 25   60 90 16 64 96

___ ___ ___ ___ ___ ___ ___ ___ ,   ___ ___ ___
18 60 12 56 81 45 96 27   16 25 96

___ ___ ___ ___   ___ ___   ___ ___ ___ ___ ___
96 64 25 72   96 12   27 96 90 45 36

___ ___   ___ ___ ___ ___   ___ ___ ___ ___ ,
90 49   108 12 81 60   64 25 6   56

                          ,

___ ___ ___   ___ ___ ___ ___   ___ ___ ___
6   49 56   56 12 49 96   84 25 96

___ ___ ___ ___   ___ ___ ___ ___ ___ ___ ___
96 64 25 72   25 27 45 6   18 25 0

Around Town

Multiply.

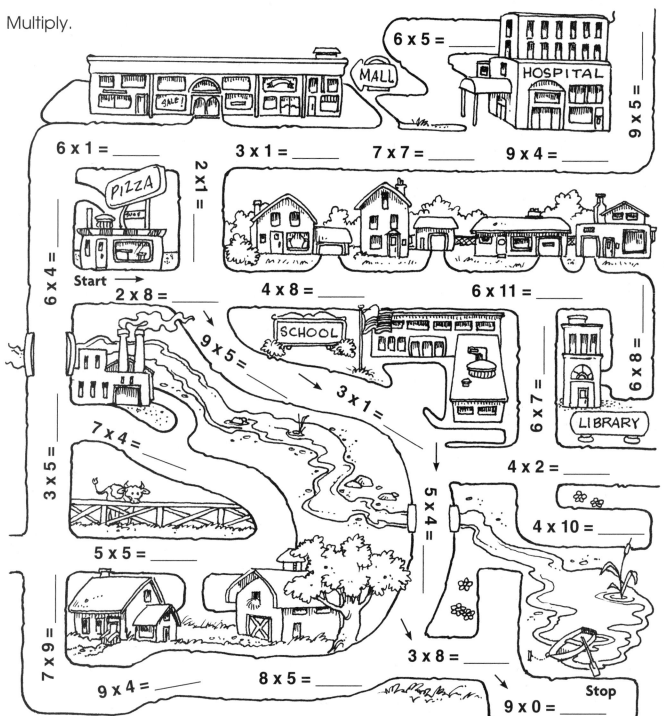

6 x 5 = _____

9 x 5 = _____

6 x 1 = _____   3 x 1 = _____   7 x 7 = _____   9 x 4 = _____

2 x 1 = _____

6 x 4 = _____

Start →

2 x 8 = _____   4 x 8 = _____   6 x 11 = _____

9 x 5 = _____

3 x 1 = _____

6 x 7 = _____

6 x 8 = _____

7 x 4 = _____

3 x 5 = _____

5 x 4 = _____

4 x 2 = _____

4 x 10 = _____

5 x 5 = _____

7 x 9 = _____

9 x 4 = _____   8 x 5 = _____

3 x 8 = _____

Stop

9 x 0 = _____

After finishing three slices of pizza at the restaurant, James walked to the pond to meet his dad. James and his dad were going to go canoeing. Add the products on the road James walked along from the pizza restaurant to the pond. Follow the arrows. What multiplication fact has a product equal to this sum?

Dot-to-Dot Multiplication

If you wanted to travel to Multiplication Island, what would be the most exciting way to get there?

To find out, multiply. Then connect the dots in order from 10 to 42.

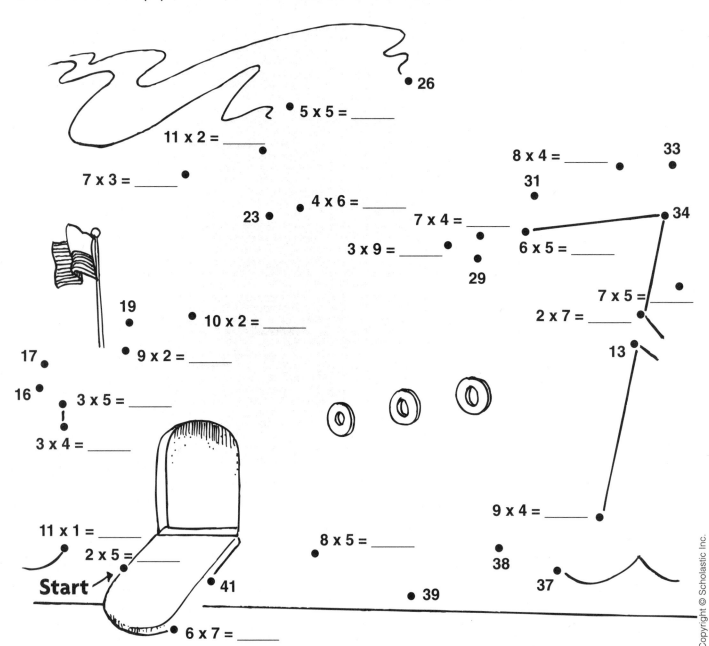

• 26

5 x 5 = _____

11 x 2 = _____

7 x 3 = _____

8 x 4 = _____   33

4 x 6 = _____   31

23 •   7 x 4 = _____   34

3 x 9 = _____   6 x 5 = _____

29

7 x 5 = _____

2 x 7 = _____

19

10 x 2 = _____   13

17   9 x 2 = _____

16   3 x 5 = _____

3 x 4 = _____

9 x 4 = _____

11 x 1 = _____   8 x 5 = _____

2 x 5 = _____   38

Start   41   39   37

6 x 7 = _____

Max and his family traveled to Multiplication Island and stayed for three days. One day Max discovered seven banana plants and five coconut palm trees. He picked six bananas from each banana plant and four coconuts from each coconut palm tree. On another piece of paper, find out how many total bananas Max picked. How many total coconuts did he pick?

Name _____

The "Ten" Flower

Multiplying by 10 is really easy! Multiply the factor by 1 and add a 0.

10 x 8 = _____ (Multiply 1 x 8 = 8, and add a zero. The product is 80.)
10 x 12 = _____ (Multiply 1 x 12 = 12, and add a zero. The product is 120.)

Multiply. Then color each space with a product less than 50 red. Color each space with a product greater than 70 orange. Color each space with a product equal to 50 yellow. Color all other spaces with a multiplication sentence green.

10 x 2 =
10 x 0 =
10 x 4 =
10 x 1 =
10 x 5 =
3 x 10 =
10 x 5 =
4 x 10 =
5 x 10 =
2 x 10 =
10 x 3 =
8 x 10 =
12 x 10 =
10 x 12 =
8 x 10 =
0 x 10 =
1 x 10 =
10 x 8 =
5 x 10 =
10 x 10 =
10 x 6 =
10 x 11 =
9 x 10 =
10 x 6 =
6 x 10 =
10 x 9 =
10 x 10 =
11 x 10 =
7 x 10 =
7 x 10 =
6 x 10 =
10 x 7 =
10 x 6 =

Cloud Ten

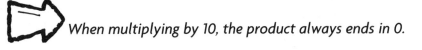 When multiplying by 10, the product always ends in 0.

Multiply.

1 x 10 = _____

10 x 9 = _____

9 x 10 = _____

7 x 10 = _____

10 x 0 = _____

3 x 10 = _____

10 x 5 = _____

10 x 8 = _____

HANG TEN!

10 x 2 = _____

10 x 4 = _____

10 x 3 = _____

10 x 10 = _____

8 x 10 = _____

WAY COOL!

6 x 10 = _____

0 x 10 = _____

4 x 10 = _____

10 x 7 = _____

COOL!

11 x 10 = _____

10 x 1 = _____

10 x 12 = _____

10 x 10 = _____

10 x 11 = _____

2 x 10 = _____

12 x 10 = _____

5 x 10 = _____

10 x 6 = _____

 Every morning Miranda chose her favorite ten clouds in the sky. She especially liked clouds that looked like animals. At the end of one week, how many clouds did Miranda choose altogether?

Eleven! Eleven!

When multiplying the factor 11 by a number from 1 to 9, double the number to find the product.

Examples: 11 x 5 = 55   11 x 7 = 77

Look at each multiplication sentence. If the product is correct, circle it. If the product is incorrect, cross it out and write the correct product above it.

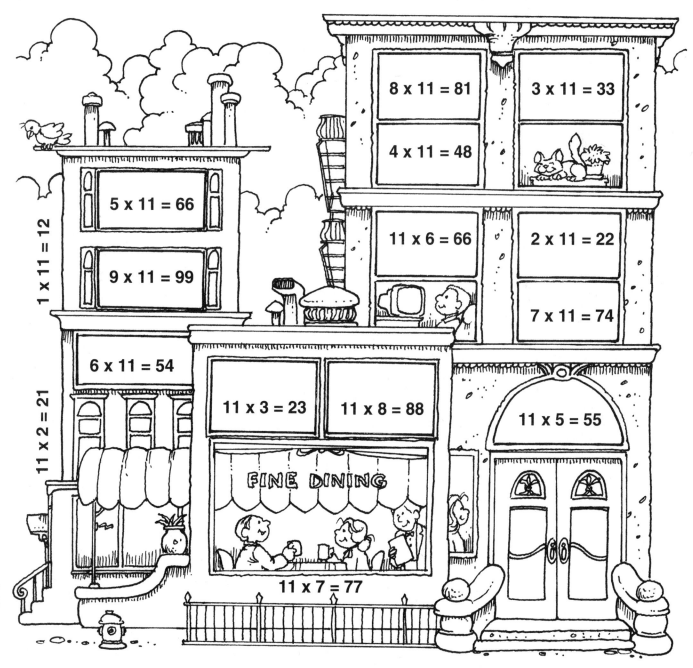

8 x 11 = 81

3 x 11 = 33

4 x 11 = 48

5 x 11 = 66

11 x 6 = 66

2 x 11 = 22

9 x 11 = 99

7 x 11 = 74

1 x 11 = 12

6 x 11 = 54

11 x 3 = 23

11 x 8 = 88

11 x 5 = 55

11 x 2 = 21

FINE DINING

11 x 7 = 77

11 x 4 = 44   11 x 9 = 88   11 x 1 = 11

The "Tuffys"

The rest of the multiplication facts with a factor of 11 are: 11 x 0 = 0, 11 x 10 = 110, 11 x 11 = 121, and 11 x 12 = 132. Since you cannot just double the number being multiplied by 11, these are the "tuffys."

Multiply. If the multiplication sentence is a "tuffy," color the space blue. If it is a double, color the space yellow.

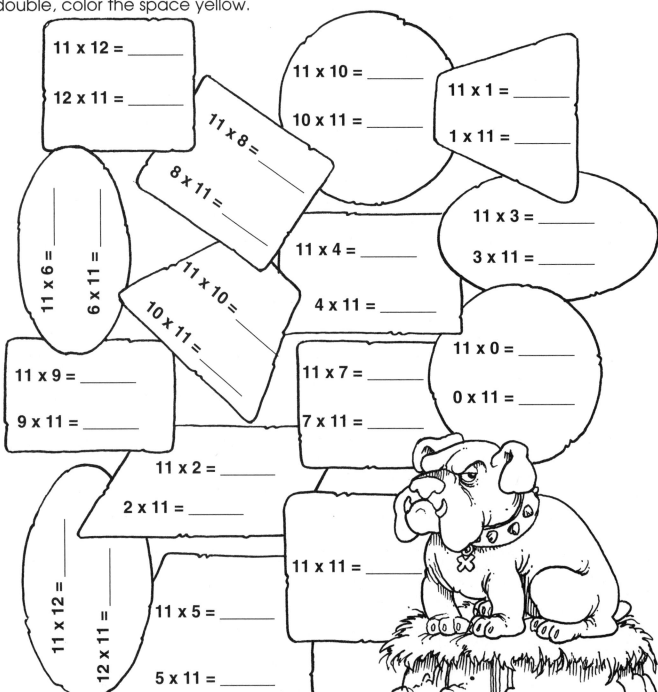

11 x 12 = _____

12 x 11 = _____

11 x 10 = _____

10 x 11 = _____

11 x 1 = _____

1 x 11 = _____

11 x 8 = _____

8 x 11 = _____

11 x 3 = _____

3 x 11 = _____

11 x 4 = _____

4 x 11 = _____

11 x 6 = _____

6 x 11 = _____

11 x 10 = _____

10 x 11 = _____

11 x 0 = _____

0 x 11 = _____

11 x 9 = _____

9 x 11 = _____

11 x 7 = _____

7 x 11 = _____

11 x 2 = _____

2 x 11 = _____

11 x 12 = _____

12 x 11 = _____

11 x 11 = _____

11 x 5 = _____

5 x 11 = _____

Searching for Facts of Twelve

Multiply. Then circle the number word for each product in the puzzle. The words will go across, down, and diagonally.

A. 12 x 0 = _____   12 x 4 = _____   12 x 7 = _____   12 x 10 = _____

B. 12 x 1 = _____   12 x 5 = _____   12 x 8 = _____   12 x 11 = _____

C. 12 x 2 = _____   12 x 6 = _____   12 x 9 = _____   12 x 12 = _____

D. 12 x 3 = _____

```
O N E H U N D R E D T H I R T Y – T W O
R N F X W F R Q R I P D B Q E H B H O P
E Q E Z T O O U C Z G S C I O D M I A F
O N E H U N D R E D E I G H T R T R T O
Y W I L U I G U N N R X D W L E N H Y R
M T E N V N H W T H I R T Y – S I X G T
E W F O N E D U K D L E A T R E B C Y Y
T E G N E T R R O H U T T D X E E O P –
Y N O H S Y H H E I G H T Y – F O U R E
P T I U E – G V B D H I D W N Q J N T I
L Y W N X S L I E E T R J F E I G H Y G
D – T H V I O W S N Y W J K F L B T N H
A F T S I X T Y O E T X E K O L V C K T
K O U G E U Y R E N I Y O N R Y V E T Z
T U R H U V D D Z S E V E N T Y – T W O
F R E S O N E F T E T Y F I X Y A Y E C
A N T F V U R N O R R M S O U T Y I R J
O N E H U N D R E D F O R T Y – F O U R
```

 Ramone went searching for rocks for his collection. He found 12 different kinds of rocks. If he found 5 rocks of each kind, how many rocks in all did he find?

Thinking Thoughts of Twelve

Write a multiplication fact in each box using 12 as a factor for the product on each wastebasket. Use a different sentence for each product.

A.

 84  0

B.

 96  132  72  144  36

C.

 12  60  84  108  48

D.

 120  48  96  132  24

 Elizabeth wrote 12 different multiplication sentences on each of 6 different pieces of paper. After solving all the problems, she discovered 5 of the problems had the same product. On another piece of paper, show how many multiplication sentences Elizabeth wrote in all. Then write 5 multiplication sentences with the same product.

There Are No Obstacles Too Big for You!

Use a stopwatch to time how long it takes to multiply around the obstacle course.

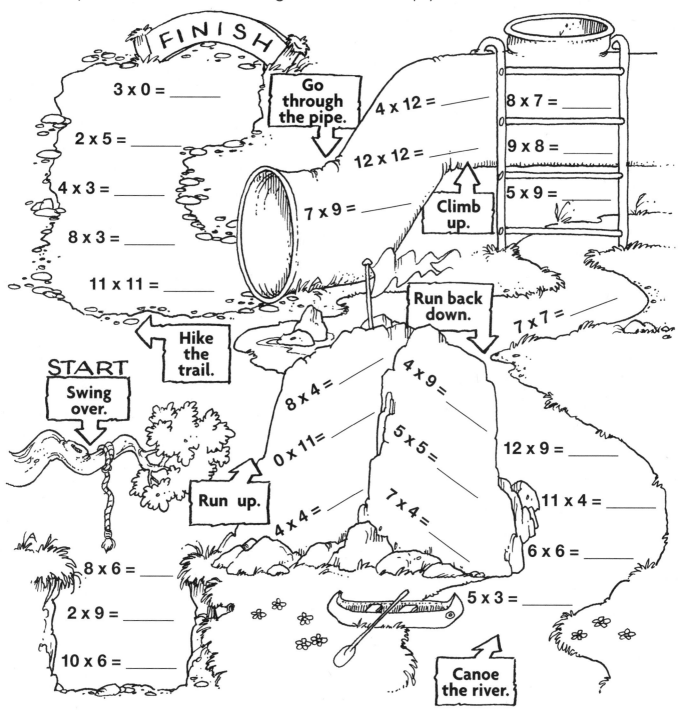

FINISH

3 x 0 = _____

2 x 5 = _____

4 x 3 = _____

8 x 3 = _____

11 x 11 = _____

Go through the pipe.

4 x 12 = _____

12 x 12 = _____

7 x 9 = _____

8 x 7 = _____

9 x 8 = _____

5 x 9 = _____

Climb up.

Run back down.

7 x 7 = _____

Hike the trail.

START

Swing over.

8 x 4 = _____

0 x 11 = _____

4 x 4 = _____

Run up.

4 x 9 = _____

5 x 5 = _____

7 x 4 = _____

12 x 9 = _____

11 x 4 = _____

6 x 6 = _____

5 x 3 = _____

8 x 6 = _____

2 x 9 = _____

10 x 6 = _____

Canoe the river.

In the morning, four students completed the obstacle course. In the afternoon, five students completed the same course. If each student completed the course seven times, how many times altogether was the course completed?

Name _____

Multiplication Success

Why are multiplicationists so successful?

To find out, multiply. Then use the code to write the letter of each multiplication sentence on the blank above its product.

A.	10 x 10 =	**G.**	3 x 1 =	**N.**	12 x 8 =	**S.**	6 x 9 =
B.	6 x 7 =	**H.**	9 x 9 =	**O.**	6 x 6 =	**T.**	6 x 0 =
C.	5 x 6 =	**I.**	8 x 9 =	**P.**	11 x 12 =	**U.**	5 x 8 =
E.	7 x 7 =	**L.**	12 x 2 =	**Q.**	8 x 8 =	**V.**	7 x 3 =
F.	3 x 9 =	**M.**	3 x 6 =	**R.**	4 x 5 =	**Y.**	2 x 8 =

___ ___ ___ ___ ___   ___ ___ ___ ___ ___ ___ ___
49 21 49 20 16   132 20 36 42 24 49 18

___ ___ ___ ___   ___ ___ ___ ___ ___ ___ ___ ___ ___
 0 81 49 16   49 96 30 36 40 96   0 49 20

___ ___   ___ ___ ___ ___
72 96   24 72 27 49

___ ___ ___ ___ ___ ___ ___   ___
42 49 30 36 18 49 54   100

___ ___ ___ ___ ___ ___ ___ ___ ___
30 81 100 24 24 49 96   3 49

___ ___   ___ ___ ___ ___ ___ ___ ___ !
 0 36   30 36 96 64 40 49 20

Page 4
A. 2 x 3 = 6; B. 3 x 3 = 9;
C. 4 x 2 = 8; D. 3 x 5 = 15;
E. 1 x 3 = 3; F. 4 x 3 = 12;
G. 2 x 6 = 12; H. 3 x 4 = 12;
I. 3 x 6 = 18; J. 5 x 3 = 15;
K. 5 x 1 = 5; L. 7 x 2 = 14;
Check array. 24 people

Page 5
A. 2 x 4 = 8; B. 3 x 3 = 9;
C. 3 x 5 = 15; D. 4 x 3 = 12;
E. 4 x 1 = 4; F. 6 x 3 = 18;
G. 8 x 2 = 16; H. 6 x 4 = 24;
I. 2 x 6 = 12; J. 8 x 3 = 24;
K. 3 x 6 = 18; L. 4 x 5 = 20;
M. 2 x 2 = 4; N. 6 x 1 = 6;
O. 5 x 4 = 20; P. 7 x 2 = 14
Check diagram.
35 hamburgers

Page 6
A. 3 x 5 = 15; B. 4 x 6 = 24;
C. 2 x 8 = 16; D. 4 x 2 = 8;
E. 3 x 7 = 21; F. 4 x 4 = 16;
G. 3 x 9 = 27; H. 5 x 5 = 25;
I. 5 x 3 = 15; J. 4 x 10 = 40;
K. 5 x 1 = 5; L. 3 x 11 = 33;
M. 4 x 8 = 32; N. 4 x 0 = 0;
O. 4 x 12 = 48; P. 4 x 9 = 36;
24 items

Page 7
A. 4 x 4 = 16; B. 7 x 2 = 14;
C. 2 x 5 = 10; D. 4 x 5 = 20;
E. 3 x 7 = 21; F. 6 x 2 = 12;
G. 8 x 3 = 24; H. 9 x 1 = 9

Page 8
A. 2 x 6 = 12; B. 8 x 4 = 32;
C. 9 x 3 = 27; D. 7 x 3 = 21; E.
9 x 5 = 45; F. 7 x 6 = 42;
G. 4 x 7 = 28; H. 12 x 3 = 36; I.
8 x 9 = 72; J. 5 x 6 = 30;
K. 10 x 4 = 40; L. 7 x 9 = 63; M.
11 x 2 = 22; N. 11 x 12 = 132

Page 9
A. 2 x 3 = 6, 3 x 2 = 6, 6 ÷ 2 =
3; 6 ÷ 3 = 2; B. 2 x 8 = 16, 8 x
2 = 16, 16 ÷ 2 = 8, 16 ÷ 8 = 2;
C. 4 x 5 = 20, 5 x 4 = 20, 20
÷ 4 = 5, 20 ÷ 5 = 4; D. 3 x 5 =
15, 5 x 3 = 15, 15 ÷ 3 = 5, 15
÷ 5 = 3; E. 3 x 9 = 27, 9 x 3 =
27, 27 ÷ 3 = 9, 27 ÷ 9 = 3;
F. 3 x 12 = 36, 12 x 3 = 36,
36 ÷ 3 = 12, 36 ÷ 12 = 3;
G. 5 x 6 = 30, 6 x 5 = 30, 30
÷ 5 = 6, 30 ÷ 6 = 5; H. 6 x 7 =
42, 7 x 6 = 42, 42 ÷ 6 = 7, 42
÷ 7 = 6; I. 4 x 8 = 32, 8 x 4 =
32, 32 ÷ 4 = 8, 32 ÷ 8 = 4; 33
÷ 3 = 11 marbles; 3, 11, 33

Page 10
Check grids. A. (3, 4), 12;
B. (2, 4), 8; C. (3, 2), 6;
D. (1, 5), 5; E. (5, 2), 10;
F. (3, 6), 18; 32 square
inches

Page 11
A. 15, 18, 21, 24, 27; B. 20, 24,
28, 32, 36; C. 5, 6, 7, 8, 9; D. 28,
35, 42, 49, 56; E. 40, 50, 60, 70,
80; F. 9, 36, 45, 54, 63; G. 18,
24, 36, 42, 48; H. 11, 33, 55, 66;
I. 20, 25, 30, 35, 40; J. 16, 32,
48, 56, 64; K. 16, 18, 20, 24, 26;
L. 12, 36, 72, 84, 96, 108;
9 miles

Page 12

Page 13
A. 6, 16, 22, 14; B. 16, 8, 4, 8;
C. 24, 10, 20, 24; D. 18, 2, 20,
14; E. 0, 12, 6, 0; F. 10, 18; G.
12, 2; H. 22, 4; Rhymes will
vary.

Page 14
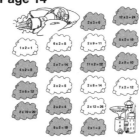

16 people

Page 15
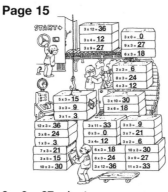

9 x 3 = 27 minutes

Page 16
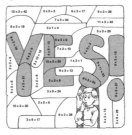

3, 30 letters

Page 17
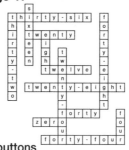

44 buttons

Page 18
A. 40; D. 16; E. 44; I. 0;
M. 8; N. 20; O. 28; R. 24; S.
12; T. 32; Y. 36; !. 48; Yes!
I adore it more and more!
Check message.

Page 19

15 scoops

Page 20
A. 10; B. 1; C. 7; D. 50; E. 12;
F. 30; G. 11; H. 15; I. 40; J. 9;
K. 5; L. 5; M. 35; N. 60; O. 4;
60 nuts

Page 21

Page 22
A. 36; B. 50; D. 16; E. 44;
G. 12; H. 18; I. 28; L. 30;
M. 32; N. 27; O. 21; P. 0;
R. 22; S. 35; T. 45; U. 24;
W. 25; Y. 48; !. 9; So that
his group would begin to
get larger and multiply!
Mr. Weaver's class has 24
students. Mrs. Moore's
class has 20 students. Mr.
Weaver's class has more
students.

Page 23
A. 22; B. 7; C. 18; D. 48;
E. 36; G. 16; H. 60; I. 12;
M. 44; O. 21; P. 40; R. 30; S.
45; T. 35; U. 28; V. 33;
Y. 24; ! 0; Because you have
the right product every time!
36 crackers

Page 24
55 seconds, 10 o'clock, 32
times, 6 hours, 18 seconds,
September 27, 24 students,
24 other students, 4 minutes,
15-minute recess, 44
minutes, 50 minutes, 7-Up,
48 times, 25 of his students,
18 days, 21 fun activities, 60
seconds, 40 problems, 60
seconds, 15 minutes, after 7

Page 25
yellow: 0 + 66 + 12 + 60 =
138; red: 18 + 30 + 36 + 6
+ 24 = 114; orange: 54 +
18 + 66 + 24 + 30 + 48 =
240; blue: 42 + 72 + 60 +
24 = 198; green: 30 + 42 +
48 + 72 + 0 = 192; purple: 36
+ 18 + 12 + 18 + 6 = 90; 48
fireworks

Page 26
A. 3 x 6 = 18; B. 6 x 5 = 30;
C. 9 x 6 = 54; D. 11 x 6 = 66;
E. 6 x 2 = 12; F. 4 x 6 = 24;
G. 8 x 6 = 48; H. 7 x 6 = 42;
I. 6 x 12 = 72; J. 0 x 6 = 0;
K. 1 x 6 = 6; L. 10 x 6 = 60;
M. 30, ★ x ✓ = ☺≋; N. 42,
✓ x ⅄ = ⇦⊠; O. 54, ✓ x ♦
= ★⇦; P. 18, ✓ x ☺ = ●⅃;
Q. 48, ✓ x ⅃ = ⇦ ⅃; R. 36,
✓ x ✓ = ☺✓; S. 72, ⌘ x ✓ =
⅄⊠; T. 60, ✓ x ☺ = ✓≋; 72
flower symbols

Page 27

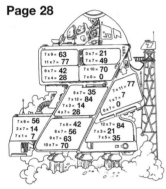

42 months; no

Page 28

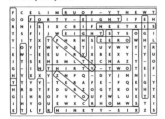

49 times

Page 29

32, 56, 72, 48, 24, 64, 96, 80, 16, 0, 24, 88, 8, 32, 56, 48, 16, 0, 88, 40, 80, 40, 96, 8, 72; 96 seconds

Page 30

A. 16, 32, 32, 80; B. 0, 40, 48, 72; C. 8, 24, 16, 88;
D. 8, 96, 24, 48; E. 40, 64, 0;
F. 72, 56, 64

Page 31

The other exception is 0 x 9 = 0.
A. one hundred eight, 12 x 9 = 108; B. ninety-nine, 11 x 9 = 99;
C. ninety, 10 x 9 = 90;
D. eighty-one, 9 x 9 = 81;
E. seventy-two, 8 x 9 = 72;
F. sixty-three, 7 x 9 = 63;
G. fifty-four, 6 x 9 = 54;
H. forty-five, 5 x 9 = 45;
I. thirty-six, 4 x 9 = 36; The facts are in descending order.; 3 x 9 = 27; 2 x 9 = 18; 1 x 9 = 9; 0 x 9 = 0

Page 32

108 pieces

Page 33

28 tiles

Page 34

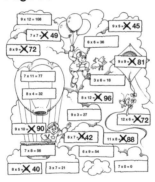

Page 35

27 words, 24 times, 24 people, January 20, 28 television stations, 9-minute interview, 72 dog bones, 16 days, 36 million people, 60 air-conditioned doghouses, 30 people, 5 highly trained tummy rubbers, 21 very famous people, 56 days, 90 barking friends

Page 36

A. 6; C. 45; D. 56; E. 25;
F. 24; G. 16; H. 64; I. 90;
K. 36; L. 84; M. 72; N. 49;
O. 12; P. 18; R. 60; S. 27;
T. 96; U. 81; Y. 108; !. 0; Find the right products, get them to stick in your head, and don't let them escape!

Page 37

The sum is 108.
12 x 9 or 9 x 12

Page 38

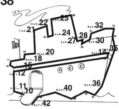

42 bananas, 20 coconuts

Page 39

Red: 10 x 4 = 40, 10 x 1 = 10, 4 x 10 = 40, 10 x 3 = 30, 1 x 10 = 10, 0 x 10 = 0, 2 x 10 = 20, 3 x 10 = 30, 10 x 0 = 0, 10 x 2 = 20; Orange: 8 x 10 = 80, 10 x 12 = 120, 8 x 10 = 80, 10 x 10 = 100, 9 x 10 = 90, 11 x 10 = 110, 10 x 10 = 100, 10 x 9 = 90, 10 x 11 = 110, 10 x 8 = 80, 12 x 10 = 120; Yellow: 10 x 5 = 50, 5 x 10 = 50, 10 x 5 = 50, 5 x 10 = 50; Green: 10 x 6 = 60, 7 x 10 = 70, 6 x 10 = 60, 10 x 7 = 70, 10 x 6 = 60, 6 x 10 = 60, 7 x 10 = 70, 10 x 6 = 60

Page 40

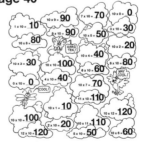

70 clouds

Page 41

Page 42

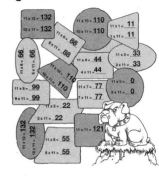

Page 43

A. 0, 48, 84, 120; B. 12, 60, 96, 121; C. 24, 72, 108, 144; D. 36

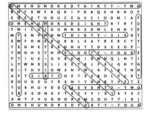

60 rocks

Page 44

A. 12 x 7 = 84, 12 x 0 = 0; B. 12 x 8 = 96, 12 x 11 = 132, 12 x 6 = 72, 12 x 12 = 144, 12 x 3 = 36; C. 12 x 1 = 12, 12 x 5 = 60, 7 x 12 = 84, 12 x 9 = 108, 12 x 4 = 48; D. 12 x 10 = 120, 4 x 12 = 48, 8 x 12 = 96, 11 x 12 = 132, 2 x 12 = 24
72 sentences; Answers will vary.

Page 45

9 x 7 = 63

Page 46

A. 100; B. 42; C. 30; E. 49; F. 27; G. 3; H. 81; I. 72; L. 24; M. 18; N. 96; O. 36; P. 132; Q. 64; R. 20; S. 54; T. 0;
U. 40; V. 21; Y. 16

Every problem they encounter in life becomes a challenge to conquer!